P9-EIE-265

Extreme Birds

Extreme

Birds

The World's Most Extraordinary and Bizarre Birds

Dominic Couzens

FIREFLY BOOKS

A FIREFLY BOOK

First printing

Publisher Cataloging-in-Publication Data (U.S.)
Couzens, Dominic.
Extreme birds : the world's most extraordinary and bizarre birds / Dominic Couzens.
[288] p. : col. photos. ; cm.
Includes index.
Summary: Birds are among the most popular and fascinating of all living things. They amaze us with their variety, stun us with their beauty, and intrigue us with their ingenuity. Packed with astonishing facts and absorbing detail, this lavishly illustrated book brings together the extremes of the bird world, from biggest and fastest to the smelliest and smartest.
ISBN-13: 978-155407-423-5
ISBN-10: 1-55407-423-1
1. Birds. 2. Exotic birds. I. Title.
598 dc22 QL676.C689 2008

Library and Archives Canada Cataloguing in Publication

Couzens, Dominic
Extreme birds : the world's most extraordinary and bizarre birds / Dominic Couzens.
Includes index.
ISBN-13: 978-1-55407-423-5
ISBN-10: 1-55407-423-1
1. Birds. I. Title.
QL673.C69 2008 598 C2008-901761-7

Published in the United States by
Firefly Books (U.S.) Inc.
P.O. Box 1338, Ellicott Station
Buffalo, New York 14205

Published in Canada by
Firefly Books Ltd.
66 Leek Crescent
Richmond Hill, Ontario L4B 1H1

Developed by HarperCollins Publishers Ltd.
77-85 Fulham Palace Road
London W6 8JB

For Harper Collins:
Editorial Director: Jilly MacLeod
Editor: Kate Parker
Design and project management: Tall Tree Ltd.
Production: Keeley Everitt

Cover design: Erin R. Holmes

This book was printed using papers that are natural and renewable and recyclable products made from wood grown in sustainable forests. The manufacturing processes conform to the environmental regulations of the country of origin.

Color reproduction by Colourscan, Singapore
Printed and bound in China by Leo Paper

To Matthew and Daniel

Contents

Introduction

Few groups of animals are as visible or abundant as birds. It is difficult to go outside without seeing at least a few, even if they are only pigeons or sparrows. Because of this, birds are perhaps the best known and appreciated of all wildlife.

Many avian record holders are both famous and obvious. The largest mammal might be hidden away under the sea, and the largest snake might keep to itself, but you cannot miss an ostrich wandering over the African savanna. It can't miss you, either, because its eyes are five times as big as yours, the largest of any vertebrate. Meanwhile, the bird with the largest wingspan, the wandering albatross, is a household name and a symbol of the destruction we have wrought in the oceans, while the feats of the fast-flying peregrine falcon and the deep-diving emperor penguin are well known to many enthusiasts.

However, if you delve deeper into the world of birds, you will find some astonishing feats that are not so well known. There is a bird, for example, that can sleep for 100 days, and another that can fly without stopping for a minimum of four years. There are those that can hear in three dimensions and spot a rodent from a mile away, and others that can fly to heights of 30,000 feet (9,000 m) without any side effects. Still others use their mouths as thermometers. Despite the fact that birds are much less physically and physiologically variable than many other animals—as a result of the constraints placed upon them by the need to fly—they still manage to take their body plan and abilities to every possible extreme.

Once we begin to look into the behavior of birds we find even more surprises. Here the "extremes" might not be so obvious, but the ways in which birds live and solve their problems are no less remarkable. Take the Arctic owls that cache the bodies of rodents for later consumption and then defrost them by sitting on them as they would eggs, or the small African parrot that carries its nest material in the feathers of its back, leaving its wings free for flight. The behaviors of birds—those small nuances of difference that have arisen to give a competitive advantage—are a mine of intrigue and astonishment.

Nowhere is this more obvious than during reproduction. Birds, it seems, will go to any lengths to get their genes passed on, be it by rape, deception or parasitism—and sometimes by all three within the same species. Yet while the young may be cosseted inside nests with exceptional insulating properties, they may also be summarily abandoned when conditions for breeding go awry. Young birds don't just sit there and accept what comes to them either; some deliberately kill their siblings, and others may fire lazy parents. Acts of desperate survival are everywhere.

No book of this kind is possible without its source material, so I must thank a group of people whose labors are sometimes overlooked—the researchers. It is they who put in the hard hours of effort and inquiry that may ultimately translate into a single sentence on the page of a book. There is no record without someone to measure it, and no discovery of previously unknown behavior without someone in the field to look and wonder. This book is really a tribute to the researchers' efforts.

It has become fashionable in recent times for every wildlife book to make a plea about conservation, but in this case repetition is apposite, and I make no apology for raking over the obvious once again. A number of the species in this book, including the wandering albatross, the Andean condor, the hooded grebe and the aquatic warbler, have low populations and are close to extinction. They are not included here for their rarity, of course, but for some extreme of lifestyle or behavior. The very fact, however, that every bird book records characteristics that are in danger of being lost is merely symptomatic of how carelessly we have treated our world. I hope that, in its small way, *Extreme Birds* will help fight against this trend and spread a little delight in the feathered creatures with which we share this planet.

Dominic Couzens
Dorset, England, April 2008

Extreme

Form

Widest wingspan · Longest legs · Biggest eyes · Smallest bird · Deadliest enemy · Most effective camouflage · Longest toes · Strongest claws · Biggest mouth · Smelliest bird · Oddest bill · Longest bill · Best energy saver · Most poisonous bird · Snuggest underwear · Ugliest looks · Lightest bones · Heaviest flier · Worst flier · Best submarine · Most flexible mouth · Oddest skin · Longest tongue · Most variable feathers · Heaviest testes · Longest penis · Sexiest tail · Biggest belly · Classiest colors · Whitest bird · Most variation in a species · Best grooming aids · Most useful esophagus

Widest wingspan

NAME **wandering albatross** *Diomedea exulans*
LOCATION southern oceans
ATTRIBUTE longest wings of any bird

A magnificent wandering albatross rides the stormy winds around the island group of South Georgia in the south Atlantic Ocean. With its long, straight wings it can cope with all but the most violent gales, translating the raw energy of tempestuous winds into masterful, effortless flight.

Albatross, of which there are about 20 species, are among the most exciting birds in the world to see, and the wandering albatross is the biggest of all of them. It holds the record for the longest wingspan of any bird, reliably measured at 12 feet (3.6 m; the norm being 8 feet–11½ feet/ 2.5–3.5 m), although informed speculation puts some individuals in excess of 13 feet (4 m).

The wandering albatross looks so powerful that you may be surprised to learn that if it has to flap its wings more than a few times in succession it quickly becomes exhausted. It is a big, heavy bird, and the wings are too narrow to carry its weight in still conditions (technically, it is said to have a high wing loading). However, if a wind of over 11 mph (18 km/h) is blowing, the sheer length of the wings comes into play in positive fashion, generating lift from wing tip to wing tip and almost negligible drag. Hence most albatross are found in the southern oceans, where winds tear around the earth unhindered.

The method used by albatross to travel vast distances with a small expenditure of energy is known as dynamic soaring. The birds glide along the trough of a wave and then suddenly turn into the wind, which scoops them up to a height of 33–50 feet (10–15 m). They then drift down toward the next trough and repeat the process for as long as they need.

A lesser flamingo takes off while the rest of the flock remains stiltlike in the water. Looking at this image, it should be of little surprise to find out that flamingos have the longest legs of any bird relative to its size. A flamingo stands up to 5 feet (1.5 m) tall, while its legs can be up to 27 inches (68 cm) long, accounting for almost half the height of the bird.

Why, though, should flamingos have such long legs? It certainly isn't to help them run, as it is for the ostrich. Being water birds with partly webbed feet, they would soon trip over if they tried to run. The advantage of their long legs is that they help them to wade in deeper water than any competitors. Because flamingos obtain all their food in the same way, by filtering water of microorganisms, they need to be able to work in water of almost any depth. Indeed, occasionally they also swim.

One of the reasons that the legs look so long is that they are absolutely bare, right up to the belly. This is an adaptation to the flamingos' habitat. Flamingos tend to feed in extremely alkaline or saline water, where very few other birds could exist. Some also live near hot springs, where they can wade in water of temperatures up to 154°F (68°C). Such conditions would cause unacceptable wear and tear on the feathers.

Longest legs

NAME **lesser flamingo** *Phoenicopterus minor*
LOCATION Africa
ATTRIBUTE proportionally the longest legs of any bird

Biggest eyes

NAME **ostrich** *Struthio camelus*
LOCATION sub-Saharan Africa
ATTRIBUTE largest eyes of any bird

When ostriches look at you it's hard to avoid their gaze. That's probably because of their rather sinister, long, snakelike neck and serious, unflinching expression. Then again, it could be because they have the largest eyes of any bird in the world.

In fact, this is not the only impressive statistic about an ostrich's eyes. At 2 inches (5 cm) in diameter, measured front to back from the center of the cornea to the retina, they are five times bigger than the human eye and beat that of any land animal; only the mightiest squids in the sea have larger eyes. Furthermore, in a neat comparison of largest to littlest, the eye of an ostrich is about as big as the smallest hummingbird! Altogether it is a remarkable organ.

The ostrich needs large eyes for its terrestrial lifestyle, sharing the savanna as it does with an alarming army of fearsome predators. The bird is famed for its running prowess, reaching speeds of 45 mph (70 km/h) when pressed, which is fast enough to escape most predators. But it has to see them first. Standing up to 9 feet (2.75 m) tall, it enjoys an excellent view over the grassland and bush. Meanwhile, the high number of photoreceptor cells in its eye, combined with the sheer size of the image from the lens, means that the ostrich can see in phenomenal detail. Indeed, its eye is at the size limit of usefulness—any larger and diffraction effects would begin to distort the image.

Smallest bird

NAME	**bee hummingbird**
	Mellisuga helenae
LOCATION	Cuba
ATTRIBUTE	smallest of all birds

The 300 or so species of hummingbird are famous for their diminutive size, so it is not surprising that one of them holds the record for being the smallest bird in the world. That accolade is generally given to the male bee hummingbird, a species from Cuba that is a mere 2¼ inches (5.7 cm) long, but there are plenty of birds that are not far behind. Some of the woodstars (*Chaetocercus* spp.) are only 2½–2¾ inches (6–7 cm) long, and the reddish hermit (*Phaethornis ruber*) rivals the bee hummingbird in lightness, tipping the scales at less than 1/14 ounce (2 g).

For now, however, the bee hummer is officially the smallest, until someone discovers a new hummingbird that has been hitherto overlooked as a flying insect. It could happen—bee hummers really do look like large flying insects, and some species of hummingbird, including the bumblebee hummingbird (*Atthis heloisa*), imitate the flight style of bees in order to avoid eviction from blooms by more aggressive, territorial hummingbirds.

Being the smallest bird in the world also affords the bee hummingbird other records. It is presumed also to have the smallest nest of any bird, measuring only ¾ inch (2 cm) in both diameter and depth. Within this nest it lays what are probably the smallest eggs, measuring just ½ inch (12.5 mm) long by ⅓ inch (8.5 mm) wide and less than half the weight of a paper clip. Indeed, it would take 3,000 bee hummingbird eggs to equal the weight of the world's largest living bird's egg, that of the ostrich.

Deadliest enemy

NAME **southern cassowary**
Casuarius casuarius
LOCATION rain forests of New Guinea and northeast Australia
ATTRIBUTE stabbing and disemboweling people

In April 1926, near the town of Mossman in northeast Queensland, Australia, a group of boys went out hunting cassowaries for fun. As they chased a bird through the rain forest, one of them, a 16-year-old name Phillip McClean, tripped over a branch and fell to the ground. In that brief moment the pursuer became the pursued. The cassowary turned on the boy and slashed him with its central claw, slicing open his jugular vein. The unfortunate McClean thus became the first, and so far only, authenticated fatality from a wild cassowary attack in Australia. He is also one of the very few people ever to have been directly killed by any bird, anywhere in the world.

Cassowaries, of which there are three known species, are large, flightless rain-forest dwellers, the third largest of all living birds after the ostrich and the emu. As tall as an average person, they weigh up to 130 pounds (60 kg), can run at 30 mph (50 km/h) and, according to some reports, can leap 5 feet (1.5 m) into the air. Normally peaceful birds, they rarely attack people except under extreme provocation.

They are well equipped to defend themselves. On top of their head is a thick casque that they use to butt into their enemies, which proves effective against most assailants. However, it's the claws that are truly deadly. A cassowary has three toes per foot, each one bearing a claw. The central claw grows up to 5 inches (12 cm) long and is as sharp as a dagger. Lashing out with it, a bird can easily slice through flesh and disembowel someone. Since 1990 alone, there have been six serious attacks on humans recorded, several of them upon zookeepers, one of which was fatal.

An adult willow ptarmigan takes refuge in a snow burrow, confident that its white plumage will protect it. It is a perfect example of camouflage in a species that, being plump and tasty, is sought out with some vigor by a range of tundra carnivores.

A high proportion of the world's bird species exhibit camouflage of one type or another, but the ptarmigans are exceptional in that their plumage changes up to three times a year to match the seasonal habitat. A white winter coat is suitable until the snow melts, but then it would become a liability, making the birds *more* obvious rather than less. Thus in the spring, prior to breeding, ptarmigans switch plumage. They begin to acquire some reddish brown feathers to match the appearance of the vegetation as it emerges from the snow. At first they are blotchy, as if the white was melting from their feathers, but by midsummer the transformation is complete, and they have become rich brown almost all over. Their cloaking is different, but it is as perfect for the time of year as their winter coat.

After breeding, ptarmigans then have two molts in quick succession. At first they change hue to a grayer, more modest version of their summer plumage, as the colors around them also grow tired and faded. Then, later on, as the tundra succumbs once again to the grip of the long Arctic winter, the changes are reflected in the patterns of the birds' feathers. White blotches appear on the plumage, which grow and coalesce until, once again, the ptarmigans are barely visible against the snowy landscape.

Most effective camouflage

NAME **willow ptarmigan** *Lagopus lagopus*

LOCATION circumpolar Arctic regions

ATTRIBUTE equipped to hide at any season

Jacanas are a superb example of a family of birds adapted in a unique way to a unique habitat. They would look like any other wading birds if it wasn't for their most obvious feature—their remarkably long, spidery toes. No other birds' toes splay out as far in proportion to their size. On the other hand, no other birds live out their entire lives on the delicately unstable world of floating lily pads.

The mechanics of the matter are simple enough. The longest toes on the largest comb-crested jacanas extend 8 inches (20 cm), and each foot can cover a surface area of about 45 square inches (300 sq. cm). This simply spreads out the bird's weight so that it doesn't sink when it is walking over floating vegetation.

It's an all-or-nothing adaptation. All aspects of their lives, including courtship display and breeding, take place on the jacanas' literal equivalent of thin ice. This places particular emphasis on the nest, which invariably floats on the surface too. Most jacana nests are small piles of vegetation; the male builds several at a time, and when the birds copulate, it is possible that the female monitors how well the nest copes with the weight of two birds before choosing to lay her eggs in a particular one.

Longest toes

NAME **comb-crested jacana** *Irediparra gallinacea*
LOCATION Australia
ATTRIBUTE extended toes

Strongest claws

NAME **harpy eagle** *Harpia harpyja*
LOCATION Central and South America
ATTRIBUTE very strong and long hind claws

The magnificent harpy eagle is the largest predatory bird in the world. Some vultures are larger, but they are mainly carrion eaters, taking only occasional live prey. This eagle is a professional killer, terrorizing virtually every medium-sized mammal and large bird that crosses its path.

The vital statistics of this great hunter are awesome. It is more than 3 feet (1 m) in length, with a wingspan of nearly 6½ feet (2 m) and can weigh 9–20 pounds (4–9 kg), and the females are much larger and heavier than the males. Despite its size, this eagle is surprisingly agile, able to negotiate its way through thick forest and to drop rapidly onto prey from a perch. Chillingly, its hind claw is nearly 3 inches (7 cm) long and as sharp as a knife.

Not surprisingly, the harpy eagle holds the avian record for seizing and carrying off the largest prey item—a red howler monkey (*Alouatta sara*) weighing 15 pounds (6.8 kg). Apart from monkeys, harpy eagles take a variety of different creatures as prey, of which sloths are a particular favorite. The latter may constitute a significant part of the harpy's diet; it is thought that their habit of sunning themselves in the treetops in the early morning makes them especially vulnerable to predation. Harpies also catch domestic pigs and goats, young peccaries, plus armadillos, porcupines, foxes and macaws. They can also deal with snakes approaching 2 inches (5 cm) in diameter by slicing them in two.

Up until now there have been no records of harpy eagles attacking humans. But if you do find yourself alone in the Amazon rain forest, make sure you keep a good lookout up above!

Biggest mouth

NAME **tawny frogmouth** *Podargus strigoides*
LOCATION Australia
ATTRIBUTE wide gape

There are a few bird families vying for the title of having the biggest mouth for size of bird, but perhaps the frogmouths of Australasia and the Far East have the best claim, not least because of their name and the unusual way they are thought to use their gape. As the photograph demonstrates, the enormous head gives the frogmouth a most peculiar, unbalanced look, with a huge front end and a truncated posterior.

The tawny frogmouth has a gape about 2 inches (5 cm) wide and can open its mouth to about the same depth. This allows plenty of room to accommodate a variety of animal prey, including the largest insects and spiders and also, occasionally, frogs, lizards, small mammals and even birds, which are usually caught on the ground after a brief sally from an elevated perch. Most are swallowed whole, but pesky prey that tries to fight back is shaken to death or struck against a hard surface to break its bones and any further resistance. There is no easy escape from the frogmouth's heavy, sharp-sided bill.

Some observations, albeit not yet confirmed, suggest that frogmouths might use their mouths in another, less conventional way. The tawny frogmouth is sometimes observed perching with its mouth wide open for minutes on end, bar the occasional snap shut. It is thought that the mouth may exude a smelly saliva that acts as bait to attract insects. If so, this would be unique among birds.

Smelliest bird

NAME **crested auklet** *Aethia cristatella*
LOCATION Bering Sea
ATTRIBUTE strong body odor

Distinctive though the crested auklet undoubtedly is to look at, this is one species that bird-watchers can identify with their eyes closed—as long as the wind is blowing in their direction. For the crested auklet, together with its close relative the whiskered auklet (*Aethia pygmaea*), is just about the smelliest bird in the world.

Colonies of these birds breed on islands in the Bering Sea. Visitors approaching by boat have been known to detect an auklet colony from as far away as 6 miles (10 km)—all because of a peculiar and very distinctive tangerinelike odor that is unique to both species.

The most interesting question about the smell is what is it for? The answer is that no one really knows, although the complex biological processes involved in developing such a unique odor must surely attest to its usefulness. The most plausible explanation is that the birds might use the smell to detect their colonies when they are returning from fishing at sea. Or perhaps they use the smell to home in on the colony where they were born, in the same way that salmon use scent to return to their natal stream.

Oddest bill

NAME **black skimmer** *Rynchops niger*
LOCATION the Americas
ATTRIBUTE peculiarly shaped bill

It's a perfect marriage of form and function. A black skimmer flies over a shallow lagoon, slicing its lower mandible through the water in an effort to detect fish by touch while lifting its upper mandible up out of the way. No other bird feeds regularly in this way, and no other bird has a bill anything like a skimmer's.

The skimmer's bill is long and large, but its most distinctive feature is that the lower mandible is longer than the upper by ½–1¼ inches (1–3 cm), creating a weird appearance, unique in a bird. But this discrepancy in jaw length is actually a by-product of the bird's feeding method, rather than at the heart of it. The act of skimming over the water surface induces a great deal of wear and tear, and from time to time the tip of the lower mandible breaks when it hits an object. Thus, the horny distal part of the bill, the rhamphotheca, grows continually, like a fingernail, at a greater rate than that of the upper mandible, producing the imbalance.

Functionally, the most extraordinary part of the bill is the lower mandible's bladelike tip, which literally slices through the water with minimal resistance. It is so narrow that it cuts through the water with little wake, and it is this part that comes in contact with fish close to the surface; farther up the bill widens considerably. Once a strike has been made, the fish is grabbed in both mandibles by a lightning-quick bend of the neck, with the head frequently facing backward as the fish is caught.

There's nothing subtle about the claim to fame of the sword-billed hummingbird, as the photograph opposite shows. The remarkably extended bill is almost as long as the bird itself. The male averages 5½ inches (14 cm), with his bill stretching to 4 inches (10 cm); the female is 5⅛ inches (13 cm) long, and her bill is a staggering 4½–4¾ inches (11–12 cm) in length. No other bird in the world comes close to matching this species in terms of ratio of bill to overall length.

It is equally obvious why the sword-bill should be so endowed. In common with dozens of other hummingbirds, its bill is perfectly shaped to match the particular bloom from which it drinks. As plants and birds have coevolved, ever more extreme blooms have provided niches for ever more extreme bill shapes; the bird beats the competition, the plant assures its pollination. As for the sword-billed hummingbird, it specializes in plants with long, hanging blooms, such as *Daturas*, passion flowers and fuchsias. These may not grow in the same profusion as other plants, but the nectar supply that each provides is rich and generous.

Of course, being so hyper-adapted causes some difficulties for this hummingbird. It must always perch and fly with its bill held up at the steep angle shown, otherwise it would become overbalanced. And preening can be a problem too: with the bill permanently out of action, the bird has to perform feather care with its feet!

Longest bill

NAME **sword-billed hummingbird** *Ensifera ensifera*
LOCATION Andes
ATTRIBUTE longest bill of any bird in relation to body length

Best energy saver

NAME **greater roadrunner**
Geococcyx californianus
LOCATION southern United States and Mexico
ATTRIBUTE heat-absorbing skin on its back

Just in case you thought that the roadrunner was merely a fictional cartoon character, the photograph opposite should convince you that it is alive and well and living outside Hollywood. It holds the record for being the fastest-running bird that can fly, reaching an impressive 15 mph (24 km/h) when pressed, although sadly it does not go "beep, beep" as it does so.

In fact, the real roadrunner is a member of the cuckoo family and is a marvelous example of a creature finely attuned to the desert environment, having a number of adaptations enabling it to cope with the notoriously variable temperatures of the desert, where the mercury soars by day and plunges by night. To meet these demands the roadrunner has resorted to acting like a reptile. Although it is warm blooded, when the temperature falls after dark the roadrunner begins to turn down its own thermostat, dropping its metabolic rate and its body temperature by a few degrees, and cutting down on energy costs. It plunges into a slightly torpid state.

In the morning it is important for the roadrunner to warm up quickly, and this it does in a highly unusual way—by using its own solar panel. On its back it has a patch of feathered skin heavily pigmented with melanin, a chemical that absorbs sunlight. The bird fluffs its feathers and the light efficiently penetrates to the skin, passing the precious heat onto the blood vessels and from there through the body. After half an hour or so of basking the bird is ready to resume running, having saved 50 percent of the energy it would have needed to rouse itself without the benefit of its very own solar-powered heating.

Most poisonous bird

NAME **hooded pitohui** *Pitohui dichrous*
LOCATION New Guinea
ATTRIBUTE poisonous feathers

In the summer of 1989 a research biologist called Jack Dumbacher was studying birds of paradise in New Guinea. The project involved catching and banding the target birds and, at the same time, releasing the many other species caught in the mist nets alongside them. Among these birds were hooded pitohuis, common birds of the forest.

One day Dumbacher was attempting to release yet another collateral pitohui when the bird ungratefully pecked and scratched him. The field-worker put his bleeding fingers in his mouth and immediately felt a curious numbing sensation. He recognized the feeling as being caused by a toxin, but it was not until he licked one of the bird's feathers that he realized that these, and not some local plant, were the source of the poison. He had stumbled upon the very first recorded instance of toxicity in a bird's plumage. Despite the fact that pitohuis had been known to science for hundreds of years, with dozens of specimens in museums around the world, nobody had noticed the phenomenon before.

The toxin in thc pitohui's plumage is the same neurotoxin found on the skin of poisonous frogs in South America. Although there is enough poison on a pitohui's skin and each of its feathers to kill several laboratory mice, the levels are still much lower than on the frogs, and it is unlikely that a predator attacking a pitohui would be killed. Nevertheless, the effects would be sufficiently unpleasant at first bite to ensure that predators would give the bird a wide berth in future. It is also possible that the toxins help to keep parasites off the pitohui's skin and feathers.

Snuggest underwear

NAME **Pallas's sandgrouse**
Syrrhaptes paradoxus
LOCATION central Asia
ATTRIBUTE allover cover of down

It can be cold up there on the windblown steppes of central Asia, where the Pallas's sandgrouse lives a mysterious life under the cover of its cryptic plumage and even more secretive habits. By contrast the temperature can be ferociously hot in the middle of the day, to the point where everything shuts down and retreats to the shade or a state of semi-torpor. The extremes of this climate represent the greatest challenge to the local fauna. It is thus that the Pallas's sandgrouse, and other sandgrouse that live in the arid regions of most of the Old World, have developed some highly unusual adaptations to cope with the great fluctuations in temperature.

The most interesting of these adaptations—one that is unique among birds—is something akin to thermal underwear. While most species of bird have down feathers arranged in tracts below the rest of their plumage, with gaps between each tract, sandgrouse have an unbroken layer of black furry down. Furthermore, the contour feathers that form the exterior of the bird are particularly downy—or "plumulaceous"—at the base. Nothing creeps through this protective layer, not even on the legs and toes, which are feathered rather than bare.

The thermal covering is ideal for temperature regulation, but it is not the only item in the sandgrouse's armory. These birds also have a greater capacity for cooling down than most other birds, achieving it by water loss through evaporation from the skin.

The portrait opposite shows a Fischer's turaco, which you will doubtless admire as a handsome and colorful bird. It would be a travesty if you didn't, because its feathers happen to be suffused with some of the rarest pigments in the entire animal kingdom. Effectively this bird is a model, and it's wearing designer plumage.

The two special pigments are called turacin and turacoverdin, named after the turaco family (Musophagidae)—a small group of 23 species found only in sub-Saharan Africa, of which the Fischer's turaco is one. Both pigments are copper based, and, so far, they have not been found in any animal other than turacos. Indeed, not all turacos have them: a couple of species seem to be lacking them.

Turacin is a red pigment that is mainly found on the wings, although on this Fischer's turaco it can also be seen to adorn the crest and nape. Turacoverdin is a green pigment found throughout the body, and its intensity is related to the habitat of the relevant species: Fischer's turaco, for example, is a forest bird and has it in abundance. Even more interestingly, this is the only green pigment synthesized by any bird. All other green colors, from the wondrous iridescence of hummingbirds to the plainer green of warblers or finches, arise through structural modifications of the feathers that cause light to be refracted unequally.

Young turacos don't manage to acquire full adult colors until they are about a year old. It seems that the required amount of copper upon which the pigments depend takes that long to accumulate from the birds' diet. The turacin found on a single bird can apparently yield about $^{1}/_{3,600}$ ounce (8 mg) of copper—and that's without taking the turacoverdin into account.

Classiest colors

NAME **Fischer's turaco** *Tauraco fischeri*
LOCATION coastal East Africa
ATTRIBUTE astoundingly rare pigments in its plumage

Whitest bird

NAME **ivory gull** *Pagophila eburnea*
LOCATION high Arctic
ATTRIBUTE entirely white plumage

An adult ivory gull flies past, barely visible against the low light of the Arctic sky. Its ethereal plumage is unusual: only a handful of birds in the world are completely white. Most have at least a few darker patches or stains here and there.

It is perhaps not surprising that the ivory gull is an exception, for it spends virtually all its life amid the pack ice of the extreme north, catching fish and feeding on carrion and waste, including dead sea mammals and their feces. It is a relatively small gull, and no doubt the plumage helps to conceal it from predators, especially when it is standing on the snow and ice. However, the color of the underparts also serves an additional purpose. In common with most gulls, ivory gulls take fish, often by dipping down in flight to the water below. Carefully conducted experiments have shown that white plumage conceals avian predators against the sky, not allowing the fish to detect the danger.

Concealment is not always the purpose behind white plumage however. A few other white birds, such as egrets and cockatoos, have white plumage to make themselves more conspicuous, not less. This enables such birds to advertise their presence to other members of the same species, either to gather together or, in the case of egrets, to indicate that a territory is filled. In fact, ivory gulls sometimes gather into flocks, and on such occasions it is even possible that their multipurpose whiteness helps them draw together.

Most variation in a species

NAME **ruff** *Philomachus pugnax*
LOCATION northern Eurasia
ATTRIBUTE greatest variation among individual males

These two ruffs might look as though they are sharing a secret, but in fact they are fierce rivals. You could say they are diametrically opposed, in both motivation and in the color of their plumage.

The ruff has an unusual breeding system in which males and females meet only for copulation. The meeting places are known as leks, and most leks are attended by 10 or so males who display directly against each other in full sight of everyone. When a female visits, all the males ruffle their magnificent head and neck feathers and hope that they, rather than a rival, will be chosen as the female's mate.

The female has a lot of choice. Male ruffs don't just differ in the strength and vigor of their display; they are all individually different to look at to a degree unique among birds. The two birds here—one with white ruffs and the other with black—are but two examples of this variation. There are also brown-ruffed males and reddish males, with no two birds looking the same. From April until June each year male ruffs are the most variable in appearance of any bird in the world.

Intriguingly, the specific color of an individual ruff is related to its behavior. The dark ruffs, including those with black or brown coloration, are known as "independents"; they hold season-long territories at a specific arena and spend their time attempting to lure the visiting females. The white ruffs, on the other hand, are known as "satellites." They don't hold territories at all but move from lek to lek, and they tend to "cheat" opportunistically, copulating with females when, for example, an independent's back is turned, distracted by a rival or another female.

Why should the independents tolerate the satellites? The most plausible theory holds that the satellites' white plumage makes the lek more conspicuous and attracts more visiting females.

Most patient feeder

NAME **shoebill** *Balaeniceps rex*
LOCATION central Africa
ABILITY standing motionless for half an hour or more

You're not going to use a bill like this for anything ordinary. The aptly named shoebill of central Africa is a true specialist, feeding almost entirely on lungfish—big, sluggish fish of well-clogged, sheltered waterways. They are not easy to catch, being large and awkward to deal with, and it takes refinements of fishing technique, as well as of the bill, for the shoebill to be successful.

One of those refinements is the shoebill's extraordinary patience. In some ways its fishing mirrors the technique of herons, waiting by the waterside and eventually striking when prey comes near. But the shoebill takes the waiting much further, sometimes staying completely still for more than half an hour; a heron, and any other stealth hunter for that matter, would have given up long before that. Observers watching shoebills feeding often miss the strike, having passed into a kind of torpor themselves.

The strike, when it comes, is a real all-or-nothing affair; it is often described as a "collapse." The shoebill lurches head first at the fish, and the rest of its anatomy follows. With a bill 7½ inches (19 cm) long it scoops up a huge mouthful, frequently containing some of the lungfish's habitat as well—water, plants and all—and it may take some time before the hunter regains its balance. A lungfish constitutes an ample meal, and after feeding the shoebill can go for several days without food. In the life of this bird, it seems, a lack of impetuous hurry is the rule.

Most prolific breeder

NAME **Eurasian collared dove** *Streptopelia decaocto*
LOCATION Europe east to southeast Asia
ABILITY attempting to raise the highest number of broods in a year

A collared dove feeds its youngsters, provisioning them with a unique type of "milk" synthesized in its crop and then regurgitated. Pigeons are among the very few birds in the world (along with flamingos and the emperor penguin) to feed their young on such a product. Supremely nutritious (containing 19 percent protein and 13 percent fat), the milk helps the youngster to grow quickly and move along the collared dove's impressive production line.

Everything about a collared dove's breeding seems to be dedicated to churning out young. For one thing the eggs are not incubated for long—only 13–18 days, which is short for the size of bird. Secondly the milk enables the chicks to grow so fast that, within a couple of weeks, they are able to fly. In this culture of haste the youngsters are turned out of the nest well before they attain anything close to the weight of the adults. A third shortcut is to overlap the breeding cycle, so that, when the father may still be feeding fledged young, the female will already be incubating the next batch of two eggs.

Many people who live in temperate parts of the world are surprised when they see nests of pigeons or doves in the middle of winter. However, this is yet another unusual feature of the breeding pattern of these birds. In contrast to the majority of birds, most pigeons and doves do not exhibit a "refractory period," a kind of post-breeding hiatus in which the relevant organs regress to prevent inappropriate procreation. So there is nothing to stop them producing young all year round. Currently the collared dove holds the record for the most broods attempted in a year—nine. This is an impressive testament to the resilience and productivity of pigeons and doves as a whole.

Strangest way to cool off

NAME **wood stork** *Mycteria americana*
LOCATION warmer regions of the Americas
ABILITY cooling off using feces

A wood stork forages in an American swamp, leaning down into the water with its bill open to catch whatever comes by. This species has the fastest known reaction time of any hunting bird; if something edible—be it a fish, frog, crayfish or anything else—brushes past its bill, the mandibles snap shut within 0.025 seconds. Not much gets away and, as a result, the wood stork usually enjoys a copious diet.

In common with most members of its family (Ciconiidae), the wood stork occurs in warm climates, including the very south of the United States and northern South America. As a result it rarely gets chilly, but overheating is a common problem. Its patient, standing-still method of feeding doesn't help either, because the wood stork often works under the glare of the sun.

Storks may cool down in various ways, including seeking shade where they can, panting and ruffling their feathers to let the heat out. But it is for another form of cooling that they are most famous—or infamous. Known politely and more scientifically as "urohydrosis," this is the practice of squirting waste products, both urine and feces, onto the legs. We may recoil at the thought of such a thing, but to a stork the resultant cooling due to evaporation is a blessed relief on a hot day; it could even be described as an indulgent luxury.

Only a few other birds practice urohydrosis. The shoebill (*Balaeniceps rex*) is one, while the American vultures (Ciconiidae) constitute the others. Indeed, the shared behavior of storks and American vultures is such that these birds are thought to be close relatives.

Best architect

NAME **rock wren** *Salpinctes obsoletus*
LOCATION western North America
ABILITY using rocks in the construction of its nest

A male rock wren sings from a slab of petrified wood on a boulder slope in the American Southwest. These birds are well named, living in territories that give them access to rocks of any kind, where they feed from cracks and under boulders.

The rock wren's attachment to hard substrates is not just one of foraging. It also stretches to the construction of its nest. On the whole, the smaller birds of the world tend to deal in fine materials for furnishings—grass, moss, hair, feathers—for the simple reason that they are light, flexible and have a neutral temperature, and it is highly unusual to find any species using such bulky material as stones. But the rock wren is an exception—in a big way.

The rock wren starts by selecting a crevice of some kind for its nest site. This site is often deep and sheltered, for example, tucked well under a large rock. One might think that, with such a hard floor, a little bit of softness underfoot would be welcome but, no, the wrens use pebbles to make their foundation, one of the very few birds in the world to do so. Once this is in place, they will, somewhat grudgingly, add some grass and other soft construction material.

This is by no means the end of their labors. For reasons that are not understood, most rock wrens now add a "feature," using up to 50 substantial stones to form a "pathway" up to the nest. Each stone may add up to one-third of the bird's own weight. It's a lot of effort and must serve some purpose, but for the moment only the rock wren knows what that is.

Smartest transporter

NAME **rosy-faced lovebird** *Agapornis roseicollis*
LOCATION southwest Africa
ABILITY transporting nest material in its feathers

Members of the parrot family are not, on the whole, famous for their nest-building prowess. This is actually an understatement—many of them don't make a nest at all, simply appropriating a hole in a tree or bank and performing a few housekeeping duties. Some species dig out their own hole, while others might chew the wood at the bottom of a cavity to make the interior a bit more comfortable.

The lovebirds of Africa, however, are an exception to this rule. Not only do they build a nest—and an intricate one at that—some species actually transport their nesting material in a way that is unique among birds.

This rosy-faced lovebird is one of the better-known species. It breeds in colonies, either in rocky crevices or the eaves of buildings, or in the communal nests of weaver birds. In the former instance the female constructs a neat cup inside the cavity using long strips of bark, leaves and coarse grass. She incises the materials from plants with her strong, sharp bill, then, when she has fashioned what she needs, she tucks the material into the feathers around her rump and lower back and flies to the nest hole relatively unencumbered.

Fascinatingly, other lovebirds vary in the way they transport their material. The related black-winged lovebird (*Agapornis taranta*) tucks its load anywhere in the plumage, whereas the yellow-collared lovebird (*A. personatus*) carries it in conventional style in its bill. Why this unusual, but convenient, feather-tucking behavior has only arisen in a small group of African parrots is a mystery.

Best thermal engineer

NAME **malleefowl** *Leipoa ocellata*
LOCATION Australia
ABILITY constructing an incubator for its eggs

A malleefowl flicks some sand behind it, using its large, powerful feet. At first sight this looks like a simple case of scratching the ground for food, kicking away excess litter to reveal tasty morsels beneath.

However, the mound upon which this malleefowl is standing is in fact its nest, and the kicking away of soil is part of a complex program of temperature regulation. The malleefowl and its kin are the only birds in the world that do not warm their eggs by applying body heat. Instead, with great labor, they build their own incubator and keep their eggs viable through a combination of solar radiation and heat generated by bacteria within rotting vegetation.

To pull off this feat of external incubation is not easy. First of all the nesting malleefowl have to build their mound, which is no small task. They begin by excavating a hollow a yard (1 m) or so deep and 10–13 feet (3–4 m) in diameter. Both birds then fill the trough with leaf litter, twigs and bark, laboriously kicking their material in from a radius of 65 feet (20 m) from the nest. They then dig an egg chamber in the top and, during a brief hiatus, allow the rain to moisten the mound and give a boost to the composting of the vegetation.

Once the female has begun to lay the first of her 20 or so eggs, the bulk of the work now passes to the male. For several months he will need to regulate the temperature of the mound by adding material or removing it, either of which may take hours every day to complete. Early on the warmth is driven by the rotting process, but in the summer the sun determines the mound keeper's schedule: at daybreak he scoops out sand to let the sun's rays heat the interior, only to fill it in later for insulating purposes when the day warms up.

In order to check the temperature the male malleefowl has a special trick. He inserts his bill into the pile up to his eye, using the interior of his mouth as a thermometer. This method is sensitive enough for the male to keep the temperature at around 91°F (33°C) for months on end, warm enough, but not too warm, for the eggs to thrive.

Longest life

NAME **royal albatross** *Diomedea epomophora*
LOCATION New Zealand waters, subantarctic
ABILITY longest-living bird recorded in the wild

Seeing this royal albatross tend its chick, the generation gap is not difficult to spot. In many small species of bird nestlings will be only a year or two younger than their parents, but in long-lived birds, such as albatrosses, there are clear generations—known as "cohorts"—within the population.

This young royal has only a 30 percent chance of surviving its first year of life, but once this tricky period has been negotiated, it can look forward to growing older than almost any other wild bird. Albatrosses as a whole regularly pass 40 years of age, and the oldest recorded wild bird was a royal albatross. Nicknamed Grandma, this individual lived on Taiaroa Head, on New Zealand's South Island. It was ringed as a breeding adult in November 1937 and was last seen, together with a chick, on June 12, 1989. Assuming that it took some 10 years to reach breeding age, it must have been about 60 years old at the time.

Based on their annual mortality rate of 3–11 percent, it is predicted that some albatrosses could live to 80 years old. However, this is without taking senility into account. In common with human beings, albatrosses grow old and more vulnerable, and their death rate sharply increases after the age of 30 or more. "Grandma," clearly, was a remarkable exception.

Most remarkable immunity

NAME **Japanese white-eye** *Zosterops japonicus*
LOCATION East Asia
ABILITY high immunity to disease

The white-eye in the photograph conveys a fragile look that is typical of this family of small, nectar-feeding birds. But appearances deceive. Evidence suggests that white-eyes are tough little creatures, with a surprising immunity to some of the most deadly of all avian diseases.

Research on Hawaii has been conducted on the epidemiology of native and nonnative birds—the Japanese white-eye (*Zosterops japonica*) being one of the latter. In the 20th century both bird malaria and avian pox entered the islands from outside, carried by mosquitoes accidentally introduced by humans. These diseases decimated the local population of native birds and were at least partly responsible for the extinction of a number of the celebrated Hawaiian honeycreepers (Drepanidinae). The white-eyes, however, which had been introduced in the 1920s, seemed to have been largely unaffected, and they spread through the islands without any apparent problems. This species is now the most common forest bird on the archipelago.

Recent research has confirmed the white-eye family's high levels of immunity. A study of mosquito-bite rates in Japanese white-eyes has shown that these birds seem to be experts in not getting bitten; furthermore, they are harder to infect than native birds, showing very low levels of the disease-causing pathogen in the blood. Similarly, research in New Zealand on the silvereye (*Z. lateralis*), a closely related species, has also shown that these birds can carry bird malaria and bird pox without being severely affected. So perhaps there is a family-wide resistance in these most unlikely of combatants on the battlefield of deadly diseases.

Largest food store

NAME **acorn woodpecker** *Melanerpes formicivorus*
LOCATION western North America south to Colombia
BEHAVIOR hoarding tens of thousands of acorns

It's one thing to make holes in trees, but this woodpecker seems to be overdoing things a bit. The species in this picture, however, is an acorn woodpecker, a bird renowned for its unusual hoarding behavior. In the northern part of its range it depends heavily on acorns; these may constitute 50–60 percent of its annual diet, and a much higher proportion in winter.

Acorns, of course, are a highly seasonal crop, so the woodpecker harvests them in fall and stores them away. The holes in this tree, as you can see, are used as small deposit boxes. The nuts, which may also include almonds, walnuts and pecans, are firmly wedged in to keep out thieves, those without strong bills, who might be tempted to help themselves when the woodpecker isn't looking.

The so-called "granaries," which may contain 50,000 holes drilled over the generations, are, however, a considerable draw to rivals. So, in order to protect a large and vital resource, acorn woodpeckers form groups that live together on a permanent basis, sharing the granary stores. During the breeding season, the groups, which may contain up to 12 adults, also nest collectively. Every member of the group helps to incubate the eggs and feed the young in the nest.

Most acorn woodpeckers use trees for their stores, but other wooden structures, including telegraph poles, may be used instead. This can be something of a headache for telephone companies, but for the woodpeckers the strong, smooth wood is ideal for their needs.

This distressing scene shows two Hood mockingbirds approaching an injured Nazca booby (*Sula granti*), as if lining up for the kill. In fact, the smaller birds are not as predatory as they seem, despite their long, curved bills and beady eyes. They are merely being opportunistic, attending the scene of an incident and hoping to profit from it, like ambulance-chasing lawyers.

It is not the mockingbirds that have caused the injuries. The young booby made the mistake of straying from its nest, and its horrific injuries were caused by attacks from neighboring adults. The mockingbirds, however, are certainly not there to help. Years of isolation on the island of Española (also known as Hood Island) have taught them to gather food from any source they can, blood included. They will drink the blood from this booby's wounds for as long as the injured bird is too weak to fight back.

Hood mockingbirds take blood and body fluids from other sources too. They gorge themselves on the placentas of sea lions and, most audaciously, peck the wounds of sea lion bulls wounded in battle. Human visitors are sometimes shocked to be followed by small parties of mockingbirds attracted to scratches on their legs. Drinking blood can be a risky strategy though. Animals don't take kindly to it, and there are many instances of mockingbirds being killed by the animals they are overzealously exploiting.

Most bloodthirsty bird

NAME **Hood mockingbird** *Nesomimus macdonaldi*
LOCATION Galapagos Islands
BEHAVIOR drinking blood from wounded animals

Most persistent hitchhiker

NAME **red-billed oxpecker** *Buphagus erythrorhynchus*
LOCATION eastern Africa
BEHAVIOR hitching a ride and snacking at the same time

This photograph shows a red-billed oxpecker in its natural habitat—on the head of a mammal! A member of the same family as the European starling, this strange bird is hardly ever seen away from the hides of large four-legged grazing mammals. It literally makes a living off other animals' backs.

Working on skin and fur, the red-billed oxpecker has unusual dietary preferences, verging on the vampirish. It is one of the very few birds in the world to specialize in eating ticks, which it gathers by using its bill like a pair of scissors to pry them off its host's hair or skin. If it wants a change it will also eat lice and various bloodsucking flies.

The common denominator in all these tasty morsels lies in the prey's diet—blood. The suspicion therefore arises that the oxpecker's taste is not for the protein in its prey's bodies, but the blood of its host. This is confirmed in a couple of ways. Firstly oxpeckers prefer engorged ticks to other ticks—they will eat 100 of the former a day—suggesting that the blood is the attraction. Secondly they often feed at animals' wounds, lapping at blood and body fluids without a tick in sight.

Certain animals act as preferred hosts. Zebras rarely top the poll, whereas giraffes, rhinoceroses and buffalo are very important, with warthogs a popular niche option. Not surprisingly the oxpeckers usually select large, tall animals, but certain hosts such as impalas, with a particular tendency to harbor parasites, buck the trend. Oxpeckers rarely forage on the back of elephants. Apparently these great beasts have particularly sensitive skin and cannot bear the touch of the oxpeckers' claws.

Silliest antics

NAME **kea** *Nestor notabilis*
LOCATION New Zealand
BEHAVIOR playing

The intelligence of parrots is well known, with the birds' ability to learn human sounds, for example, leading to their enormous popularity as pets and attractions at wildlife parks. Less celebrated is their capacity for play, which is well developed and perhaps reaches its pinnacle in the kea, a high-altitude species from New Zealand. In the tough environment where the keas live, adapting to the world needs to start early in life, and play is a good way of learning about relationships, foraging and survival.

Hikers in New Zealand won't necessarily be impressed. They are sometimes persecuted by gangs of keas stealing their food and fiddling around with anything left unattended. Keas have been known to steal cups and mugs and throw them downhill like mischievous adolescents. To make matters worse, they have been recorded sliding down the sides of tents early in the morning, waking the inhabitants up.

Sliding is, in fact, a common pastime. In the cold mountains where they live, keas regularly tumble down snowy slopes, sometimes on their back. They also slide down windshields and roofs and, in captivity, have been known to make snowballs. Such play is usually conducted in groups of all ages, and the members frequently play fight and just mess around. Other antics include lying with their feet in the air and kicking at individuals on top of them and even swimming on their backs.

In a gesture of togetherness, a male superb fairywren treats its mate to a spot of mutual preening. The closeness of the two birds is not just physical; fairywrens keep their mates for life and reside with them in the same territory all year round. The familial bliss is added to by the youngsters, which often remain with their parents for a year or more. Encountering fairywrens in the wild, it is normal for an observer to see a small group of them.

Nevertheless the relationship between this pair could be described as "open" because, although the two birds live together, their sexual relationships frequently stray outside the pair-bond. Both sexes can be promiscuous; indeed, the superb fairywren is one of the least "faithful" of all birds to its primary relationship. In a study it was found that only 24 percent of eggs laid by a female fairywren were fathered by her social mate. The reasons for such a high rate of extra-pair copulation are unclear.

The behavior of the female is particularly intriguing. During her fertile period in the breeding season she makes regular predawn visits to copulate with various males. The male is equally accommodating to females that might visit him during the same time of day.

The seed of this twilight activity is sown in the months before breeding begins, when male fairywrens make regular visits to neighboring territories, displaying their wares. They are often decidedly indiscreet, displaying to the incumbent females even when the territory-holding male is present. Hilariously, they will even bring gifts of flowers in their bills, and these are always yellow. It is on such meetings, it would appear, that the extra-pair copulations are planned. So the female fairywren's gesture of conjugal fidelity seen here conceals a rather more promiscuous reality.

Craziest suitor

NAME **superb fairywren** *Malurus cyaneus*
LOCATION southeastern Australia
BEHAVIOR male bringing flowers to his "mistress"

A fully paid-up member of the fraternity of small brown birds, the aquatic warbler hides a surprising lifestyle beneath a deceptively demure exterior. Its pairing system is unique, involving an unusually prolonged union between male and female that could be the longest of its kind in the world.

Aquatic warblers breed only in Europe, in a scarce and declining habitat—marshes with clumps of sedge up to 32 inches (80 cm) tall with water in between. Such a habitat is exceptionally rich in insect life, a factor that is thought to have emancipated male aquatic warblers from any kind of breeding duties. The female builds the nest, incubates the eggs and rears the young on her own, while the male simply sings and seeks copulation.

Not surprisingly, with so little else to do, the males expend a great deal of effort on ensuring their paternity. They have multiple partners, and their testes are larger than usual for the size of bird, suggesting that sperm competition is fierce. They also have another extraordinary strategy to ensure that it is their genes that prevail. In the spring evenings, when the females are most fertile, males seek them out and, presumably to keep other males away, simply remain on their back for an astonishing length of time. While in most birds copulation lasts less than a second, in the aquatic warbler it can go on for half an hour.

The strategy is not as definitive as it sounds, however. Most clutches of six or so eggs are sired by two males, and some by as many as four. In such circumstances the long coupling must be a necessity for ensuring any breeding success at all.

Longest copulation

NAME **aquatic warbler** *Acrocephalus paludicola*
LOCATION Europe
BEHAVIOR perhaps the longest-lasting copulation among birds

Oddest time-share

NAME **band-rumped storm petrel** *Oceanodroma castro*
LOCATION Galapagos Islands and Azores archipelago
BEHAVIOR two populations sharing the same nesting burrows

In the bird world, nest sites tend to be sacrosanct. Where possible, a good many species return to the same site each year to breed, sometimes coming back from thousands of miles away, homing in with complete accuracy. Once established, owners of sites don't usually take kindly to challengers and, where there is conflict, it can be bloody and final. A nest site stolen is a nest site violated.

This band-rumped storm petrel is doubtless like every other bird in this respect, claiming its site vigorously every year. But what is different in this case is that—knowingly or otherwise—it shares its burrow with another breeding pair of the same species. The two sets of owners, however, do not actually meet at the burrow. This is because they have different breeding seasons; in effect, they time-share. So far, two populations of band-rumped storm petrels have been found to do this. On the Galapagos Islands one shift lays its eggs in May or June, while the other breeds from November to January. Meanwhile, on Baixo and Praia in the Azores archipelago one shift lays in the summer (June to July), the other in fall (October to December). It also appears that, in the Azores at least, individuals do not switch from one shift to another.

Indeed, in the Azores population there is a very slight difference in morphology between the hot-season and the cool-season breeders. The "cool" birds are slightly larger and heavier than the "warm" ones, and their eggs are longer and heavier. The differences are minute, and the birds are indistinguishable in the field—to people and presumably to each other. Genetic studies, however, hint that these birds may be a few steps on the way to becoming separate species.

At the same time, across the continent, other northern wheatears are arriving on the tundra of eastern Canada and Greenland, having also wintered in Africa. Theirs is not such a long journey as that taken by Alaskan birds, but they still have to cross nearly 1,900 miles (3,000 km) of ocean between Africa and Greenland. Due south of their breeding grounds lie some pleasant lands in continental South America that look ideal for wintering. They are far closer, but the wheatears ignore them too.

Of course, the northern wheatear doesn't willfully ignore the more sensible option of wintering in the New World. It just isn't programmed to take it. Instead, its behavior is thought to be a superb example of how migratory journeys were affected by the last glaciation. During the last ice age, northern wheatears were almost certainly restricted to breeding seasonally on lands within easy reach of Africa. As the ice retreated the birds gradually colonized the lands opening up, spreading west and east as well as north. They would have spread little by little, stretching their longitudinal limits year by year, yet always returning to Africa as a refuge for the winter. Eventually, of course, they would have reached a point at which they were migrating unnecessarily long distances, but this would never have occurred to them.

Nowadays, with its migration route wildly over the limit, the northern wheatear remains stoically unaware of the easier paths it could be following.

Shortest migration

NAME **Clark's nutcracker** *Nucifraga columbiana*
LOCATION mountains of western North America
BEHAVIOR making minute seasonal movements

When people think of migration, they tend to imagine long journeys that take birds north, south, east or west. But there is a host of species for whom latitudinal movement is irrelevant—it's altitude that matters.

An example is the Clark's nutcracker of western North America, which usually breeds at between 5,900 and 8,200 feet (1,800 and 2,500 m) in altitude, often in woods at the outer edge of the tree line. This bird is a particularly hardy character that can breed very early in the spring, when heavy snow is still on the ground. However, it will not necessarily see out the winter at this altitude; in fall it sometimes retreats a few hundred yards lower down. By taking the edge off the winter cold, this relatively small shift may make all the difference to the bird's chances of survival.

Can this be considered a migration? Absolutely. Just because it is measured in yards rather than miles does not diminish the migration's effectiveness. All over the world, birds accrue the same advantage from minute downhill displacements that they would gain from several hundred miles of latitudinal movement. The climate is often radically different at the bottom of a mountain range in comparison to the top.

One of the curiosities of altitudinal migration is that in fall, when most birds in the Northern Hemisphere are moving south, some altitudinal migrants move north. The water pipit (*Anthus spinoletta*) is a good example of this, breeding in the mountains of Central Europe and wintering in Britain and the Low Countries.

A common poorwill—named after its soft, sweet "poor-will" advertising call—sits motionless among the rocks in the Arizona desert, its superb camouflage making it almost impossible to spot. However, the bird's coloration needs to be exceptionally cryptic. This member of the nightjar family is famous for its ability to "hibernate," remaining motionless and therefore highly vulnerable for long periods of time. It is the only bird in the world that sees out the winter period by sleeping.

Strictly speaking, the poorwill doesn't hibernate, which would involve chemical changes in the body. It simply goes torpid, allowing its metabolic rate to drop below normal so that it uses less energy. The poorwill, however, pulls off this trick to a greater extent than any other bird and for much longer too. While many species of bird, including other nightjars, are known to become torpid overnight, or perhaps for a day or two, the poorwill can settle into this state for as long as 100 days. In so doing it is able to avoid the winter cold, which diminishes its food supply—nocturnal insects—to almost nothing.

The torpid state reduces the poorwill's various bodily functions, including heart and breathing rates, and at the same time the body temperature drops. This drop can be spectacular, from a normal 104°F (40°C) down to an extraordinary 41°F (5°C), the latter by far the lowest recorded for any species of bird. Although verified by science only recently, the sleeping habits of the poorwill have been known for generations. Indeed, the local Native American name for this bird may be translated as "the sleeping one."

Longest sleep

NAME **common poorwill** *Phalaenoptilus nuttallii*
LOCATION deserts of North America
BEHAVIOR becoming torpid for weeks on end

Most devoted ant follower

NAME **ocellated antbird**
Phaenostictus mcleannani
LOCATION Honduras south to Ecuador
BEHAVIOR following ant columns

The ocellated antbird is one of a large family of approximately 200 species of mainly brown, shade-loving birds of the South and Central American forest understory. Seeing a name like "antbird" you would think their main source of sustenance must be obvious. You would be wrong. The family is named for following ants, not for eating ants.

Ants play a significant part in the ecology of tropical forests, not least the various species of army ants that rampage along the woodland floor eating every living thing in their path. These army ant columns are permanent fixtures; each individual army moves across an area during the day and forms temporary bivouacs at night. Due to the fact that there are always columns in the vicinity of any stretch of forest, some birds have made a career out of attending the ants, making their living by picking off the many insects that flee from the hordes. Panicked by the approach of the ants, these insects become very easy to catch.

Thus it is that, each day, the ocellated antbird wakes up and seeks out a column, especially one comprising the aggressive species *Eciton burchellii*. Checking bivouac sites and previously used paths, or listening for the calls of fellow ant followers, the bird will locate a column and stay with it throughout the day, fielding crickets, cockroaches and even such formidable prey as spiders and scorpions fleeing for their lives. The ocellated antbird rarely, if ever, hunts in any other way. With such bounty so readily available it doesn't need to.

Extreme

Coziest nest

NAME **American goldfinch** *Carduelis tristis*
LOCATION North America
BEHAVIOR building a snug, watertight nest

A female American goldfinch brings in food for her young. In contrast to many small birds, this species doesn't bring in juicy caterpillars, worms or other small or soft-bodied invertebrates; instead it provides seeds regurgitated in a kind of paste. Very few songbirds feed their young on such "difficult" food.

There are several other unusual features in the domestic life of the American goldfinch. For a temperate songbird, for example, its breeding season starts exceptionally late. It is mid-July before most goldfinches start to raise young. This may be for several reasons. American goldfinches molt in the spring, which might make the multitasking load of changing feathers, nest building and laying eggs too much for the adults to bear. The birds may also be awaiting the midsummer blooming of thistles, their favorite food. At any rate, many goldfinches lay eggs as late as August, with the young fledging in October.

Another remarkable feature is the nest itself. Built mainly by the female, although the male may bring in material, it is a snug cup constructed from various plant fibers woven together exceptionally tightly with spiderwebs and plant down (especially thistledown). This nest is placed in the fork of a tree, 33 feet (10 m) up, and is so perfectly constructed that, later on when the young have left, it can hold water as reliably as any human-made receptacle.

Most eggs in a nest

NAME **ostrich** *Struthio camelus*
LOCATION sub-Saharan Africa
BEHAVIOR producing a large and variable clutch of eggs

The nest being tended by this ostrich in South Africa contains a healthy tally of eggs, but it's a long way off the record for the species. Some nests hold 20–25 eggs, and the world record is a mind-boggling 78. That's a lot of young ostriches! But not all of these will be the product of a single bird. Most nests host the eggs of at least 2–5 females, with a record 18 individuals contributing to a single clutch. You could, therefore, call these communal nests, but ostrich politics are not as egalitarian as that term implies. Not all the eggs are equally valued.

Each ostrich nest is tended by the territory-holding male and one female, who is known as the "major hen." She acquires this status by being the first female to lay an egg at the nest site. Over a period of two weeks or so the major hen lays about eight eggs in the same nest; meanwhile other females are invited to lay eggs too. These secondary females, known as "minor hens," mate promiscuously with other local males and contribute nothing to incubation anywhere, despite laying eggs in several nests.

This leads to the question: why do the male and major hen tolerate the presence of extra eggs in their nest when they get no assistance and have no genetic stake in the eggs? The answer appears to be that they use the extra eggs as a buffer against predators. If an enemy strikes the sheer numbers of eggs will reduce the chances of the major hen's eggs being destroyed. Furthermore, the major hen usually recognizes her own eggs and makes sure that they are at the center of the clutch, thus ensuring they will be incubated. The eggs from the minor hens, meanwhile, are arranged in a circle around the nest. Many will not be incubated, but some will—ensuring that there is some value in the arrangement for the minor hens as well.

Most eggs in a season

NAME **brown-headed cowbird**
Molothrus ater
LOCATION North America
BEHAVIOR laying more eggs than any other wild bird

The domestic scene in this photograph appears idyllic enough, but all is not as it seems. If you look closely you will see that one of the chicks has a yellow gape, while the rest have a pink gape. That's because the yellow-gaped chick is not the offspring of the attending wood thrush (*Hylocichla mustelina*), but is actually an interloper. During the laying of the clutch, a brown-headed cowbird stole in and laid its own egg, destroying one of its host's in the process. Now, unwittingly, the wood thrush is raising a cowbird along with her own chicks.

The very same scene is replicated in untold numbers of nests every breeding season in North America. The brown-headed cowbird is a generalized brood parasite, which means that it will lay its eggs in the nests of many different species, not just wood thrushes. In fact, in all it has been known to parasitize more than 220 different species and, of these, 144 have successfully raised chicks contracted out to them. Being a larger species than the cowbird, this wood thrush is in a way fortunate. If a cowbird lays in the nest of a species smaller than herself her young will aggressively outcompete the host youngsters, causing them all to starve to death.

Another extraordinary feature of the brown-headed cowbird's bizarre life is the fecundity of the female. With so many possible hosts to attack, she just carries on laying eggs for as long as she can. In the wild it is not uncommon for the female to lay 40 eggs in a season, and in captivity a figure of 77 has been recorded. (The common cuckoo, *Cuculus canorus*, by comparison, lays only 20.) The downside is that it causes a great deal of destruction and reduced production among the cowbird's legion of host species.

Largest extended family

NAME **noisy miner** *Manorina melanocephala*
LOCATION eastern Australia
BEHAVIOR up to 22 birds feeding the young in one nest

Of all the world's young birds, few seem to have more attention lavished upon them than the nestlings of one of eastern Australia's most common and overbearing birds, the noisy miner. The youngsters may be fed up to 50 times an hour, one of the highest rates known among birds.

Furthermore, this is very much a collective effort. An extraordinary total of 22 adults has been recorded visiting a single brood during the nestling stage to bring edible gifts, with a range of 6–21 being recorded on other occasions. All this for a clutch of only 2–4 eggs on average. These young are treated like royalty.

The extraordinary attentiveness of the adult members of this extended "family" is partly due to the fact that the noisy miner is a colonial, cooperatively breeding species. When the female is nest building, which takes about a week, she continually advertises what she is doing with a special display so that all the males in the area know what is going on and when the young might hatch from their eggs. Why they should all help her, though, is unclear. Although the birds behave promiscuously, most clutches are sired by just one father.

Noisy miners have an astonishingly complicated social structure. Males live within their specific "activity spaces," often overlapping with those of other males. Birds with overlapping spaces form long-lasting associations known as "coteries," usually comprising 10–25 birds.

Coteries make up extended "colonies," consisting of hundreds of birds, and sometimes the coteries divide into short-lived gangs known as "coalitions." It is generally the males in coteries that, together, help at a single nest—which might partially explain why so many birds are compelled to engage in this useful, though curious, type of feeding activity.

Safest nest site

NAME **great dusky swift** *Cypseloides senex*
LOCATION central South America
BEHAVIOR nesting behind a waterfall

Just below the rim of the thundering Iguassu Falls, on the border between Argentina and Brazil, four great dusky swifts appear to be toying with their lives. Wheeling inches from the roaring water, the slightest misjudgment of their flight path could cause them to strike the fast-moving surface and be washed away to certain death.

But it's an everyday risk the swifts take because, to them, the benefits of living near the waterfall greatly outweigh the dangers. Indeed, paradoxically, it is sanctuary they seek in this extreme environment. By placing their nests beside the falls, or even behind them, the swifts can keep their eggs or young uniquely safe. No predator in its right mind is likely to climb down the perilously moist rocks, bathed in permanent spray, to reach the narrow ledges upon which the swifts have built their nests. And even if a predator wanted to do so, the barrier of falling water would certainly prove impenetrable.

There are a few disadvantages, of course, to living in such an environment. The saturated conditions mean that the breeding adults have to cope with spray on their plumage when incubating or brooding. The young, on the other hand, have a generous coating of down to keep them warm. Aside from the inherent danger, the highly aerial swifts have all that they need. Luxuriant plant growth on the nearby rocks provides nesting material, and the swifts have a short commute to nearby forest to catch insects for their food.

On first seeing this picture you are probably wondering what that egg is doing there. There's no nest in sight—just an egg balancing on a branch with an ethereal-looking white seabird staring soothingly at it. Something is surely not right.

But it is. Despite the incongruity of it all, this egg is lying where it was laid, put there intentionally. What you are looking at is the nest site of the white tern, perhaps the most precarious nest site of any bird. The slightest of bumps would dislodge this egg irretrievably, causing it to tumble down to the sand below where it would probably crack. Yet the tern risks this balancing act for safety's sake, to protect its egg from ground predators.

Remarkably this site is relatively low for a white tern. It is common for the eggs to be in the canopy of trees, sometimes a dizzying 66 feet (20 m) above ground, although they are usually at about half that height. The branch seen here is, admittedly, on the narrow side. Most eggs are balanced on typically horizontal limbs that measure at least 4–5 inches (10–13 cm) wide, but only 3 inches (8 cm) has been recorded.

Terns don't leave everything to chance. They normally scratch out bark with their feet to give the egg more balance, and they will also lay the egg in a natural depression. Furthermore, the incubating adults are careful to approach the egg from behind during a changeover and to leave the site by falling backward. Even so, accidents and strong winds lead to high wastage.

The adults incubate the egg for 28–32 days and then the youngster hatches. Fitted with strong claws, a chick can hold onto a branch rather better than an egg can, but the 60 days that pass before it fledges must still be a nervous time for all.

Most precarious nest site

NAME **white tern** *Gygis alba*
LOCATION tropical islands
BEHAVIOR laying its egg on a narrow branch

Fairest chick provision

NAME **common kingfisher** *Alcedo atthis*
LOCATION much of Eurasia, Indonesia and North Africa
BEHAVIOR giving each youngster a fair chance to be fed

A common kingfisher tends its young. The kingfisher burrow might not look like an unusual scene but what goes on here is truly exceptional. Most parent birds don't instinctively look out for the welfare of every one of their chicks. The worst offenders only serve food preferentially to their firstborn, often allowing other chicks to starve, while at the majority of nests a delivery is something of a free-for-all, and the most vigorously begging nestlings receive first pick each time. A nest full of chicks is a competitive, ruthless environment.

Except, that is, among kingfishers. In the nesting chamber a rare type of fairness prevails. For their first two weeks nestling kingfishers wait to receive food in what is effectively a line. They lie down in a circle, back to back, and just one of the youngsters orients itself toward the light coming in from the nest-hole entrance. When an adult arrives with a fish only this one chick begs, and it alone is fed. The rest remain passive. Once a chick is fed it moves out of the way, and the next youngster takes its place.

This remarkable arrangement, known as the carousel system, ensures that each youngster is regularly fed. Furthermore, if a nestling tries to cheat the rest of the brood administers group justice, pecking at it and on occasion throwing it to the back of the chamber.

A male Burchell's sandgrouse visits a waterhole in Botswana. It is a daily ritual for these birds because their diet of seeds compels them to drink regularly; each dawn sees them taking commuting flights to a reliable water source. This bird appears to be taking things a step further by having a bath at the same time.

It's a very special type of bath, though, almost unique to the sandgrouse family. What this male is doing is getting his belly feathers so completely soaked that, in a few moments, he can fly back to the nest with his feathers loaded with water. His chicks are thirsty, and, once the male arrives, they will gather under his belly to have a drink of what remains on his plumage "sponge."

In order to carry out this water delivery the male sandgrouse has specially adapted feathers on his belly. The microscopic projections on the feathers' branches—the barbules—are usually coiled and lie flush against the vane (the flat plane of the feather). When soaked, however, the barbules uncoil and project at right angles to the vane, forming a bed of hairs that absorbs water by capillary action. The result is an absorption rate superior to a sponge; a male sandgrouse is estimated to be able to carry up to 1⅓ fluid ounces (40 ml) of water. Some, however, is lost to evaporation during the flight home.

Best water carrier

NAME **Burchell's sandgrouse** *Pterocles burchelli*
LOCATION southern Africa
BEHAVIOR carrying water soaked in its belly feathers

Most suspicious mate

NAME **great tit** *Parus major*
LOCATION Eurasia
BEHAVIOR escorting the female all day long

Two great tits have a skirmish at a bathing site. It is spring, and the males are bundles of nerves. Tension is boiling over.

It is the season when every male great tit must be unusually vigilant. From the period of about two weeks before laying eggs to the time, a month later, when the clutch is finally complete, a female great tit is highly liable to stray and compromise her mate's treasured paternity. Although great tits are monogamous, females regularly seek extra-pair copulation to add variety to the genetic material in the nest. It is up to their mates to stop them.

The lengths males go to to guard their mates are remarkable. Their primary weapon is song; every day throughout this period, at dawn chorus time, the male chooses to perch right outside the nest holes and sing. Once awake and active, the females come outside and, with the male ready and waiting, the sexes copulate.

During the day the male takes the same stifling approach. He keeps the female company at all times, fighting off unwelcome suitors (as seen here) and ensuring that nothing happens without his knowledge. It is during the evening, perhaps, that his paternity is most at risk because most eggs are laid at dawn. A slack moment would court disaster. So the male redoubles his efforts until, at last, the female goes to roost in the nest, still escorted by her mate.

Then, just to make sure, the male stays outside the nest, a chaperone by the door. He might wait 15 minutes before, confident of the female's sleepy state, he himself can at last retire to roost.

Strangest courtship

NAME **western bowerbird**
Chlamydera guttata
LOCATION western and central Australia
BEHAVIOR doing more than most to impress the females

Pity this poor western bowerbird—or indeed any male bowerbird. Few, if any, species in the world have to work harder to win the affections of the opposite sex. In most bird species a bit of singing and display goes a long way, but with bowerbirds that's only half the story. Bowerbirds do indeed sing and display, but they also go to astonishing lengths to make a special construction to impress the females. This construction—known as a bower and built purely for display—involves the male in hour upon hour of painstaking work followed by virtually continuous maintenance all year round. The quality of workmanship is one of the crucial factors in a female's choice of potential father to her chicks.

The bower of the western bowerbird is known as an avenue bower. It consists of two parallel stick walls, each about 14 inches (36 cm) long and 9 inches (23 cm) high and standing 6 inches (16 cm) apart. At each end the male clears away debris on the ground, replacing it with a large number of decorative objects, in this instance some green fruit; on some bowers there may be as many as 200 of these items. The designer also adds other objects, which may include bleached bones, snail shells, pebbles and sometimes human artifacts, such as glass, metal and even spent shotgun cartridges.

As if that isn't enough, the bower needs a little interior decoration as well. For this reason the bowerbird selects a juicy fruit and daubs the pulp on the inside walls, as if adding a coat of paint. But females are notoriously hard to please, and, even after years of work, some males fail to father any chicks.

Canniest false alarm

NAME **barn swallow** *Hirundo rustica*
LOCATION throughout much of the world
BEHAVIOR sounding a false alarm

Barn swallows, especially those that live in colonies, are notoriously prone to erring from the path of conjugal fidelity. Both males and females, although secured in a formalized relationship with a member of the opposite sex, nonetheless frequently find time to consort with others and thus exchange genetic material.

To a female an extra-pair copulation may be an advantage, since it offers sperm that could be of superior quality to that of her social mate. To a male, however, the idea that his mate would stray outside the pair-bond is an anathema—it would compromise his paternity. Thus male swallows regularly undertake "mate guarding." It is well named: the male follows his partner everywhere while she is fertile, ensuring that she does not have the opportunity to consort with a stranger.

Mate guarding, however, isn't easy. Sometimes the male loses sight of his mate, dramatically increasing the potential threat to his paternity. So what does he do? He gives an alarm call, warning the colony of attack from a predator. Everybody interrupts what they are doing and flies up into the air, bunching together, calling.

Except that there's no attack. It's a deliberate false alarm. In the tightly packed melee, however, it is much easier for the male to find his mate. A little deception goes a long way—especially when it keeps the female faithful.

Most notorious kidnappers

NAME **white-winged chough** *Corcorax melanorhamphos*
LOCATION eastern Australia
BEHAVIOR kidnapping the fledglings of neighbors

One of Australia's strangest birds, the white-winged chough is an inhabitant of the dry country of the east. It lives in small groups of about seven birds, which forage by walking in lines in a sinister fashion, like policemen conducting a precise search.

Each group is a close-knit, permanent association, consisting of a dominant male, several mature females and a few immature birds, the latter offspring from previous seasons. In the breeding season the group defends a communal territory, and tasks, such as constructing an unusual mud-bowl nest, brooding the chicks and bringing the nestlings food, are shared by everybody. It seems that bringing up youngsters is difficult for white-winged choughs, so much so that it requires at least four adults to complete the job effectively. Not surprisingly, recruitment into the group is painfully slow.

If the group is small, however, white-winged choughs have a way of circumventing the problem of too few helpers. They pick a fight with a neighboring group and kidnap a fledging from them. During the frequent melees between clans adults have been seen displaying at a fledgling, enticing it to swap flocks.

The kidnapped fledglings don't always remain with their "captors," but often they do, and for the long term. As a result the genetic variability within a group is enhanced, and the birds have that vital extra help on hand next time they attempt to breed.